YOUR KNOWLEDGE HAS VALUE

David Brückner

The Theory of Special Relativity

An Introduction

GRIN Verlag

Bibliografische Information der Deutschen Nationalbibliothek:

Die Deutsche Bibliothek verzeichnet diese Publikation in der Deutschen National-
bibliografie; detaillierte bibliografische Daten sind im Internet über http://dnb.d-
nb.de/ abrufbar.

Imprint:

Copyright © 2012 GRIN Verlag GmbH
Druck und Bindung: Books on Demand GmbH, Norderstedt Germany
ISBN: 978-3-656-34044-7

This book at GRIN:

http://www.grin.com/en/e-book/206888/the-theory-of-special-relativity

GRIN - Your knowledge has value

Der GRIN Verlag publiziert seit 1998 wissenschaftliche Arbeiten von Studenten, Hochschullehrern und anderen Akademikern als eBook und gedrucktes Buch. Die Verlagswebsite www.grin.com ist die ideale Plattform zur Veröffentlichung von Hausarbeiten, Abschlussarbeiten, wissenschaftlichen Aufsätzen, Dissertationen und Fachbüchern.

Visit us on the internet:

http://www.grin.com/

http://www.facebook.com/grincom

http://www.twitter.com/grin_com

The Special Theory of Relativity

David Brückner

Lancing College, West Sussex, United Kingdom

March 2012

Until the end of the nineteenth century, the simple Galilean principle of relativity was used to relate physical observations in one frame of reference to another moving relative to it. When the phenomena of electromagnetism and light where unified in Maxwell's equations, this principle was first called into question as it stood in conflict with the idea of absolute time and motion. The most famous experiment that attempted to determine the absolute motion of the earth, the Michelson-Morley experiment, will be discussed here. Subsequently, the ideas and postulates contained in Einstein's first paper on relativity will be introduced and hence the kinematic transformations based on the principles will be derived and their implications on the relativity of space and time as well as on Newtonian mechanics will be stated.

1 Galilean Invariance

The principle of relativity was first formulated by Galilei in 1661 in his book "Dialogues on two world systems" where he makes observations on the invariance of physical events on a moving boat as opposed to a stationary one [?], in other words, physical phenomena are unaffected by the choice of reference frame from which they are observed [?]. The principle assumes that there is a universal time shared by all reference frames. Considering $\mathbf{F} = m\mathbf{a}$ leads to the restriction to unaccelerated, so called inertial frames. Only then will the motion of a particle be correctly described in all reference frames [?, ?].

To test whether an equation obeys the principle of relativity, a transformation that translates the values of physical observables from one frame to another is required. Consider two frames K and K' with K' moving at velocity $\mathbf{v}$ relative to K in the x-direction as shown in Figure 1. The Galilean transformation relates the spacetime-point $(t, \mathbf{x})$ in K to the same point $(t', \mathbf{x}')$ in K', provided the systems coincided when $t = t' = 0$ in the following fashion:

$$\mathbf{x}' = \mathbf{x} - \mathbf{v}t \text{ and } t' = t. \tag{1}$$

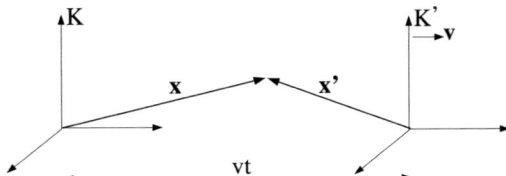

Figure 1: *Two inertial frames K and K' which moves at speed $\boldsymbol{v}$ relative to K.*

These have long been thought to be consistent with Newton's laws as experiments demonstrate that the mass and force are constant in all inertial frames, unless one is dealing with speeds close to that of light [?]. The transformation should hence yield $\mathbf{a} = \mathbf{a'}$ and indeed

$$\frac{d\mathbf{x'}}{dt'} = \frac{d\mathbf{x}}{dt} - \mathbf{v} \tag{2}$$

and

$$\mathbf{a'} = \frac{d^2\mathbf{x'}}{dt'^2} = \frac{d^2\mathbf{x}}{dt^2} = \mathbf{a}. \tag{3}$$

As this simple principle works so nicely in Newtonian mechanics, it was for long seen as being self-evident. However, when at the end of the nineteenth century the long investigations into the phenomena of light, electricity and magnetism culminated in Maxwell's equations of the electromagnetic field, first published in 1865 [?, ?], which describe these phenomena in one uniform system, these transformations were called into question as Maxwell's equations did not seem to obey the principle of relativity [?, ?]. That is, if the transformations are substituted into Maxwell's equations, they do not remain the same which would imply for example, that electrical and optical phenomena could be used to determine the absolute speed of a spaceship without looking at the surroundings which violates the principle of relativity.

One of the implication of Maxwell's equations is that light propagates at a constant speed $c \approx 3 \times 10^8$ ms^{-1}, independent of the motion of the source [?]. This brings up an interesting problem: Consider a spaceship moving with speed v and light from the rear travelling at speed c. According to (2), the speed measured in the spaceship should hence be $c - v$ and not c. It should therefore be possible to measure the absolute velocity of the earth with respect to a universal reference frame, the hypothetical aether that was supposed to pervade all space.

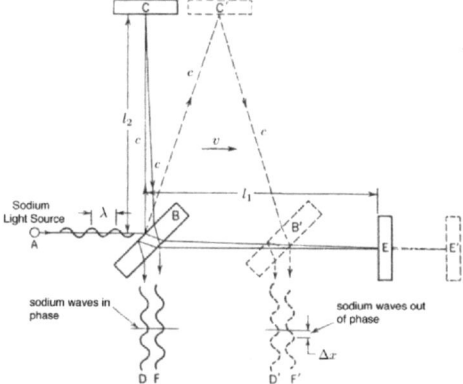

Figure 2: *Schematic representation of the Michelson-Morley experiment.*

2 The Michelson-Morley Experiment

The most famous of the many experiments that have been performed on this matter is the experiment conducted by Michelson and Morley in 1887 [?, ?] using an interferometer as shown schematically in Figure 2. Their idea was to measure the difference in the time of travel of a light beam in two perpendicular directions [?]. The apparatus consists of a sodium light source A shining monochromatic light onto a semi-silvered glass plate B wich splits the incoming beam into two beams continuing in mutually perpendicular perpendicular directions to the mirrors C and E, where they reflect back to B. On returning to B, the are joined into two superimposed beam, D and F. If the time of travel for the beams are equal, the waves will be in phase but if they differ slightly, interference will occur [?, ?].

It can be shown, that if the apparatus is at rest in the aether, no interference will occur but if it is moving at a velocity v to the right, the times should differ. First, let us calculate the time taken to travel from B to E and back. Let the time to travel forth be t_1 and back t_2. As the apparatus moves a distance vt_i during the time of travel, it can be said that

$$ct_1 = l_1 + vt_1 \text{ and } ct_2 = l_1 - vt_2.$$

Hence, the total time of horizontal travel t_h is

$$t_h = t_1 + t_2 = \frac{2l_1}{c-v} + \frac{2l_1}{c+v} = \frac{2l_1}{c(1-\frac{v^2}{c^2})}.$$

In order to find the time to go from B to C, t_3, the equations

$$(ct_3)^2 = l_2{}^2 + (vt_3)^2 \text{ and hence } t_3 = \frac{l_2}{\sqrt{c^2 - v^2}}$$

can be established. By symmetry, the distance forth and back are the same, so the vertical time of travel t_v is

$$t_v = 2t_3 = \frac{2l_2}{\sqrt{c^2 - v^2}} = \frac{2l_2}{c\sqrt{1 - \frac{v^2}{c^2}}}$$

In order to account for the possible difference in l_1 and l_2, the apparatus was turned through $90°$ to record the shift in interference [?]. Taking this into account, it was expected that there would be interference as $t_h \neq t_v$ making it possible to measure the absolute velocity of the earth. However, no interference was found – the result of the experiment was zero [?, ?]. The experiment was repeated many times and other experiments were performed but they all gave the same result [?]. The velocity of the earth through the aether could not be detected.

3   Einstein's Postulates

The question why Nature would apparently yield up no information about our motion with respect to a hypothetical fundamental frame of reference troubled the minds of some of the best physicists of the nineteenth century. Most of them took the view that the aether existed but that special mechanisms were at work that would undo every phenomenon that would permit a measurement of the absolute velocity v [?]. The first fruitful idea for such a mechanism came from Lorentz and Fitzgerald (independently) in 1892. They suggested that material bodies contract along their direction of motion through the aether and if this contraction is by a factor of $\sqrt{1 - v^2/c^2}$, the zero fringe shift follows directly [?, ?].

However, this result seemed to be too artificial, designed solely for explaining away the difficulties. The French mathematician Poincaré was the first one who suggested that there *is* such a law of nature, that it is impossible to discover an absolute velocity by experiment [?].

It was Einstein who, instead of imposing preconceived ideas on the facts, brought a grand clarity of outlook in his 1905 paper "Zur Elektrodynamik bewegter Körper". He pointed out that the analysis of motion had always been

4

based on the assumption that there was a universal, absolute time. The following quote from the paper [?] describes the starting point of his argument better than any paraphrase could:

> We need to consider that all our judgements in which time plays a part are always judgements of simultaneous events. If, for example, I say "That train arrives here at 7 o'clock," I mean something like this: "The pointing of the hand of my watch to 7 and the arrival of the train are simultaneous events."

This is of course almost trivial but Einstein goes on arguing that the case becomes problematic if it concerns the relationship between events that occur at different locations in space. Consider two observers with a clock at points A and B that can both make time-related observations in their surroundings. According to Einstein's analysis, it is not possible to make comparisons of events in A and B without further assessment as, even though there is an A-time and a B-time, no common time has been defined. This can only be done by *defining* that the time light takes to travel from A to B is equal to the time it takes to travel the way back. Consider a light beam that leaves A when the A-time reads t_A, reaches B at B-time t_B, is reflected back to A and reaches A at $\bar{t}_A$. Then, by Einstein's definition, the clocks run simultaneous if

$$t_B - t_A = \bar{t}_A - t_B. \tag{4}$$

A corollary of this definition is that the velocity of the light signal, given by

$$V = \frac{2\overline{AB}}{\bar{t}_A - t_A} \tag{5}$$

will have the same value in all circumstances, hence $V = c$ [?]. This leads to the definition of the following two postulates [?]:

Postulate 1: **All inertial frames are equivalent with respect to all the laws of physics.**

Postulate 2: **The speed of light in empty space always has the same value c.**

It is striking that a whole new dynamics can be built on these two short statements that will fully resolve the inconsistencies of theory and experiment with an exalted simplicity which could only have been discerned by a genius mind like Einstein.

4 The Lorentz Transformations

In order to test these postulates on Newton's and Maxwell's equations, a new kinematic transformation from one frame to another moving relative to it is required. These transformations were first formulated by Lorentz in 1904, a year

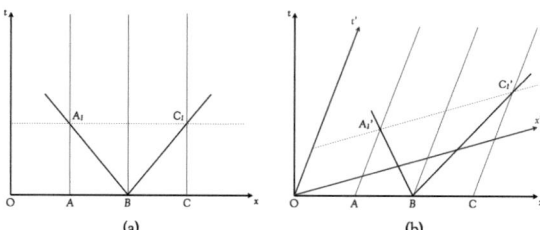

(a)　　　　　　　　　　(b)

Figure 3: *(a) Space-time diagram showing an experiment to define simultaneity at A and C which are at rest in this reference frame K. (b) Equivalent experiment with A, B and C at rest in K' which is moving relative to K as seen in K.*

before Einstein's paper, when he worked out that Maxwell's equations would remain unchanged under them [?, ?].

Here, they shall be derived from first principles (based on a method from [?]) like Einstein did independently of Lorentz. To derive the Galilean transformations (1), one only had to look at the geometry in the space dimensions as a universal time was presumed. Now space-time geometry has to be considered as, like it could be seen in Einstein's thought experiments, our judgement of time and simultaneity are a function of the particular reference frame used [?].

Consider again the inertial frames K and K' with K' moving at speed v along the x-axis relative to K. Three observers A, B and C are at rest equally spaced along the x-axis of K. Their world-lines are vertical as their spacial positions are constant. Now consider a light signal sent out at B at $t = 0$ whose world line will be $x = x_B \pm ct$. The arrival of the signal at the other observers is given by the intersection of the world lines of A and C and the world line of the signal, A_1 and C_1 (see Figure 3a). Simultaneity in the positions A and C is hence given by the line $A_1 C_1$.

Now consider A, B and C at rest in K'. Their world lines with respect to K' are now at an angle as they change position relative to K at a rate v. The arrival of the signal at A_1' and C_1' is not simultaneous in K as $A_1' C_1'$ is not parallel of the x-axis in K (see Figure 3b). As a consequence of the 1. Postulate however, A_1 and C_1 are simultaneous events in K'. This now allows us to add the coordinate axes of K': The axis of x' is the line that is parallel to $A_1 C_1$ since this is the line defining simultaneity and any line $t' = \lambda$ where λ is a constant is parallel to the x'-axis. The t'-axis is simply the world line of the origin of K'.

The point event P can therefore be expressed alternatively by x and t or

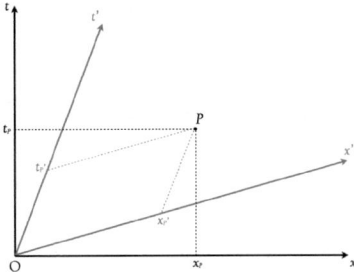

Figure 4: *The point event P as seen in K and K'.*

by x' and t' (see Figure 4). The relativity principle implies that the kinematic transformations must be bilinear, meaning that x' and t' are both linear functions of x and t and vice versa, because otherwise, a motion at uniform velocity in K would not be seen as at constant velocity in K'. This symmetry implies that:

$$x = \lambda x' + \mu t' \text{ and } x' = \lambda x - \mu t. \tag{6}$$

The velocity of K' relative to K in terms of λ and μ can be found by looking at the motion of the origins by substituting $x = 0$ into (6.1) and $x' = 0$ into (6.2). This gives $v = \mu/\lambda$.

The link between the two frames is the light beam as it is the only motion invariant under motion. A signal sent of at O will be described as $x = ct$ and $x' = ct'$ in K and K' respectively. By substitution in (6)

$$ct = (\lambda c + \mu)t' \text{ and } ct' = (\lambda c - \mu)t$$

can be obtained. By elimination of t and t' and using $\mu = \lambda v$ one finds

$$c^2 = \lambda^2(c^2 - v^2)$$

and hence

$$\lambda = \frac{1}{\sqrt{1 - v^2/c^2}} =: \gamma(v). \tag{7}$$

The required transformation can now be found by using $\lambda = \gamma(v)$ and again $\mu = \lambda v$ in equation (6):

$$x = \gamma(x' + vt') \text{ and } x' = \gamma(x - vt) \tag{8}$$

7

which reduce to the Galilean transformation (1) at low speeds, when $v/c \to 0$. By simple rearranging, the transformation in time can now also be found:

$$t = \gamma(t' + vx'/c^2) \text{ and } t' = \gamma(t - vx/c^2). \tag{9}$$

It follows directly from the 1. Postulate that the quantities that measure distance transverse to the direction of motion are equivalent in all frames because otherwise, there would be ways of detecting absolute motion and displacement [?, ?], so:

$$y' = y \text{ and } z' = z. \tag{10}$$

The equations (8), (9) and (10) together form the Lorentz transformations which express any K' coordinates in terms of K and vice versa.

These transformations have consequences that go against the assumptions of classical physics. From the point of view of an observer at rest, a clock in motion will have go slower than his and a rod in motion will have a shorter lenght than the same rod at rest [?]. These corollaries are called time dilation and length contraction.

The time dilation can be derived by considering two events (x_0, t_1) and (x_0, t_2) in K. The time coordinates in K', moving at v relative to K, are $t'_1 = \gamma(t_1 - vx_0/c^2)$ and $t'_2 = \gamma(t_2 - vx_0/c^2)$. The lapse of time in K' in terms of the observed difference in K is therefore

$$t'_2 - t'_1 = \gamma(t_2 - t_1).$$

This is often written with $t_2 - t_1 = \tau_0$ and $t'_2 - t'_1 = \tau$ [?] as

$$\tau = \gamma\tau_0 = \frac{\tau_0}{\sqrt{1 - v^2/c^2}}. \tag{11}$$

The length contraction is easily found by considering a body whose two ends are marked by x_1 and x_2 in K and consequentially has length $l_0 = x_2 - x_1$. In K' this distance at the same instance is judged as $l = x'_2 - x'_1$ where, by transformation, $x_1 = \gamma(x'_1 + vt')$ and $x_2 = \gamma(x'_2 + vt')$. Thus,

$$x_2 - x_1 = \gamma(x'_2 - x'_1)$$

and hence,

$$l = l_0\gamma^{-1} = l_0\sqrt{1 - v^2/c^2} \tag{12}$$

which is exactly the contraction proposed by Lorentz and Fitzgerald.

We can conclude, that the time passes more slowly for a moving observer than a stationary one and that the measured length of a material body is less in moving frame than in the rest frame [?, ?]. This has been verified by many experiments and observations for instance by the elongation of the half-life of muons travelling at a speed close to c [?].

8

5 Relativistic Dynamics

The transformations derived from Einstein's theory work in Maxwell's equations and have simplified the theoretical structure of electrodynamics and optics [?, ?] whilst Newtonian mechanics had to be modified before it came in line with special relativity. It turns out that all that has to be done is to replace the mass in Newtons second law

$$\mathbf{F} = \frac{d(m\mathbf{v})}{dt}$$

by the expression

$$m = \gamma m_0 = \frac{m_0}{\sqrt{1 - v^2/c^2}}, \tag{13}$$

where m_0 is the rest mass of the body [?]. This can be deduced by looking at a simple scenario [?]: Consider a particle of mass m_0 travelling at constant speed v in the x-direction. Relative to a stationary observer, its momentum is given by

$$\mathbf{p} = m_0 v = \frac{\Delta \mathbf{x}}{\Delta t'}.$$

Relative to an observer moving with the particle however, the momentum is given by

$$\mathbf{p} = m_0 \frac{\Delta \mathbf{x}}{\Delta t} = m_0 \frac{\Delta \mathbf{x}}{\Delta t'} \frac{\Delta t'}{\Delta t} = \gamma m_0 \mathbf{v}$$

By expanding the expression $m = \gamma m_0$ and multiplying both sides by c^2 one obtains

$$mc^2 = m_0 c^2 + \frac{1}{2} m_0 v^2 + \frac{3}{8} m_0 v^4 / c^2 + \dots \tag{14}$$

in which mc^2 is the total energy of the body, $m_0 c^2$ an intrinsic energy called the rest energy and the terms added to the rest energy represent the increase of mass arisen from kinetic energy [?]. Einstein came to the result $E = mc^2$ by the following Gedankenexperiment [?, ?].

Suppose a burst of light with energy E is emitted from one end of a box of mass M and length l (see Figure 5). Combining $E = hf$ and $\lambda = h/p$, the momentum of the radiation is given by E/c. Therefore, the box recoils with speed

$$v = -\frac{E}{Mc}.$$

Having travelled freely for a time Δt, the radiation hits the other end of the box and brings it to rest again after it went a distance Δx. If $v \ll c$, $\Delta t \approx l/c$ and thus

$$\Delta x = v \Delta t = -\frac{El}{Mc^2}. \tag{15}$$

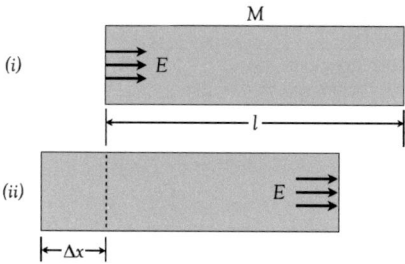

Figure 5: *Einstein's box.*

As this is an isolated system, Einstein postulates, in order not to violate the conservation of mass-energy, that the light has carried an equivalent of a mass m, so that

$$ml + M\Delta x = 0. \tag{16}$$

The combination of (15) and (16) yields the grand result

$$E = mc^2. \tag{17}$$

The following proof shows that this is consistent with $m = \gamma m_0$ [?]:

$$\frac{dE}{dt} = \mathbf{F} \cdot \mathbf{v} \tag{18}$$

which, by substitution of (17) and Newton II, gives

$$\frac{d(mc^2)}{dt} = \mathbf{v} \cdot \frac{d(m\mathbf{v})}{dt}. \tag{19}$$

To make the integration easier, both sides can be multiplied by $2m$:

$$\int dt \, 2mc^2 \frac{dm}{dt} = \int dt \, 2mv \frac{d(mv)}{dt}$$

which allows us to write

$$m^2 c^2 = m^2 v^2 + C$$

and by using the boundary conditions $v = 0$ and $m = m_0$,

$$m^2 c^2 = m^2 v^2 + m_0^2 c^2 \text{ that is } m^2 (1 - v^2/c^2) = m_0^2 \tag{20}$$

from which the result

$$m = \frac{m_0}{\sqrt{1 - v^2/c^2}} = \gamma m_0$$

is found.

References

[1] G. Galilei, *Dialogues on two world systems.* Translated by T. Salusbury, 1661, available at `http://archimedes.mpiwg-berlin.mpg.de/cgi-bin/toc/toc.cgi?step=thumb&dir=galil_syste_065_en_1661`, accessed 22. February 2012.

[2] R. P. Feynman, *Six Not-So-Easy Pieces.* Penguin Books, 1st ed., 1999.

[3] E. Taylor, J. Wheeler, *Spacetime Physics: Introduction to Special Relativity.* 2nd ed., 2001.

[4] A. P. French. *Special Relativity.* M.I.T. Introductory physics series. Chapman and Hall, 1st ed., 1968.

[5] H. Lipson, *The Great Experiments in Physics.* Oliver and Boyd, 1st ed., 1968.

[6] B. Ridley, *Time, space and things.* Penguin Books, 2nd ed., 1984.

[7] A. Einstein, *Relativity - The Special and General Theory.* Translated by R. W. Lawson, Routledge, 1st ed., 1960.

[8] W. Rindler, *Introduction to Special Relativity.* Oxford Science Publications, 1991.

[9] M. Jarrell, *The Special Theory of Relativity.* 2001. Available at `http://www.phys.lsu.edu/~jarrell/COURSES/ELECTRODYNAMICS/Chap11/chap11.pdf`, accessed 20. February 2012.

[10] A. Einstein, "Zur Elektrodynamik bewegter Körper," *Annalen der Physik,* vol. 322, issue 10, pp. 891-921, 1905. Available at `http://onlinelibrary.wiley.com/doi/10.1002/andp.19053221004/pdf`, accessed 24. February 2012.

[11] A. Einstein, "Das Prinzip von der Erhaltung der Schwerpunktsbewegung und die Trgheit der Energie," *Annalen der Physik,* vol. 325, issue 8, pp. 627-633, 1906. Available at `http://wikilivres.info/wiki/Das_Prinzip_von_der_Erhaltung_der_Schwerpunktsbewegung_und_die_Trgheit_der_Energie`, accessed 26. February 2012.

[12] A. Waser, *On the Notation of Maxwell's Field Equations.* 2000. Available at `http://www.zpenergy.com/downloads/Orig_maxwell_equations.pdf`, accessed 23. February 2012.

[13] L. Campbell and W. Garnett, *The Life of James Clerk Maxwell*. Macmillan and Co., 1882. Available at http://www.sonnetsoftware.com/bio/maxbio.pdf, accessed 23. February 2012.

[14] S. Sinha, *Poincaré and the Special Theory of Relativity*. 2000. Available at http://www.ias.ac.in/resonance/Feb2000/pdf/Feb2000p12-15.pdf, accessed 25. February 2012.

[15] C. Bishop, *Particle Physics*. John Murray Ltd., 2002.

Picture Credits

Figure 1: http://www.phys.lsu.edu/~jarrell/COURSES/ELECTRODYNAMICS/Chap11/chap11.pdf
Figure 2: http://www.relativitycalculator.com/images/Albert_Michelson_Part_II/michelson-morleyexperiment.png
Figure 3, 4, 5: A. P. French. *Special Relativity*. M.I.T. Introductory physics series. Chapman and Hall, 1st ed., 1968.